AMAZING SPACE

THE SUN

BY MATT LILLEY

CREATIVE EDUCATION • CREATIVE PAPERBACKS

Published by Creative Education and Creative Paperbacks
P.O. Box 227, Mankato, Minnesota 56002
Creative Education and Creative Paperbacks
are imprints of The Creative Company
www.thecreativecompany.us

Design by The Design Lab
Art direction by Graham Morgan
Edited by Joe Tischler and Ana Brauer

Photographs by Getty Images/MARK GARLICK, 6; NASA/GSFC/SDO, 18, GSFC/Solar Dynamics Observatory, 2, SDO, HMI, cover, SDO, HMI, 14; Science Source/Detlev van Ravenswaay, 17; Unsplash/Jongsun Lee, 21, Melissa van Niekerk, 5, NASA, 10, Ravi Sharma, 20; Wikimedia Commons/ESA/Hubble, 7, J. Falck, 13, Museo Civico di Modena, 16, NASA, 9, 11, NASA/Johns Hopkins APL/Steve Gribben, 22–23, NASA/SDO (AIA), 1

Library of Congress Cataloging-in-Publication Data
Names: Lilley, Matt, author.
Title: The sun / Matt Lilley.
Other titles: Amazing space (Creative Education, Inc. (Mankato, Minn.))
Description: Mankato, Minnesota : Creative Education and Creative Paperbacks, [2025] | Series: Amazing space | Includes bibliographical references and index. | Audience: Ages 6–9 | Audience: Grades 2–3 |
Summary: "Learn about the composition and formation of our solar system's largest star, the Sun. This astronomy title for elementary-aged kids includes captions, on-page definitions, a space science feature, additional STEM resources, and an index"— Provided by publisher.
Identifiers: LCCN 2024018581 (print) | LCCN 2024018582 (ebook) | ISBN 9798889892380 (library binding) | ISBN 9781682776049 (paperback) | ISBN 9798889893493 (ebook)
Subjects: LCSH: Sun—Juvenile literature.
Classification: LCC QB521.5 .L54 2025 (print) | LCC QB521.5 (ebook) | DDC 523.7—dc23/eng/20240509
LC record available at https://lccn.loc.gov/2024018581
LC ebook record available at https://lccn.loc.gov/

Printed in China

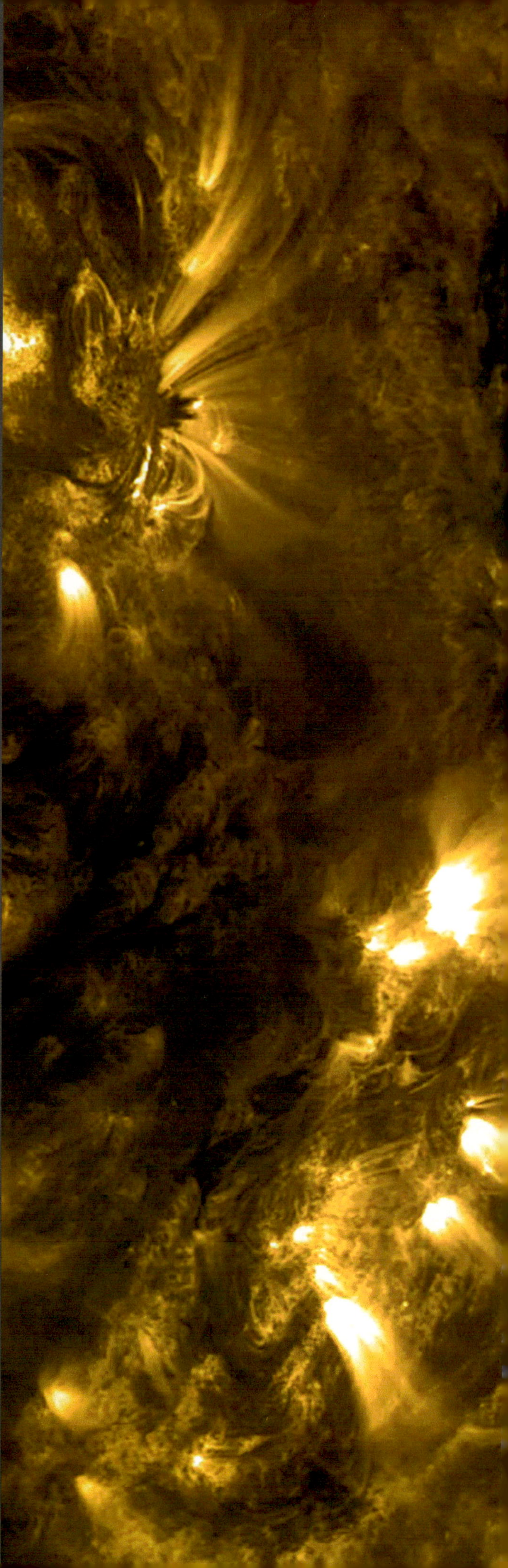

Table of Contents

The Sun is a star. It is the closest star to Earth. That's why it looks bigger than other stars. It is a kind of star that scientists call a "yellow dwarf."

The Sun is the only star in our solar system.

The next closest star to Earth is about 23.5 trillion miles (37.8 trillion km) away.

The Sun is at the center of the solar system. Every planet in the solar system revolves around it. The Sun is about 93 million miles (150 million kilometers) away from Earth. Sunlight takes about eight minutes to reach Earth.

revolve to move around something in a loop

The Sun takes up the space of over a million Earths.

The Sun is the largest object in the solar system. It has the **mass** of about 330,000 Earths. The Earth **rotates** once a day. The Sun also rotates. It takes about 27 days.

rotate to spin

mass the amount of matter or substance that makes up an object

The Latin word for Sun is Sol.

The Sun is mostly made up of hydrogen and helium gas. The gases are heated up into plasma. The surface of the Sun is about 10,000 degrees Fahrenheit (5,538 degrees Celsius). The center is much hotter.

plasma a substance that is similar to a gas but that can carry electricity

Nicolaus Copernicus was a **polymath**. *He studied math and* **astronomy**.

In the 1500s, people thought that the Sun orbited Earth. This was wrong. Nicolaus Copernicus used math to understand how the planets move. He proved that the planets orbit the Sun.

polymath a person with a wide range of knowledge

astronomy the study of everything in space

COLAV
OPERNICVS.

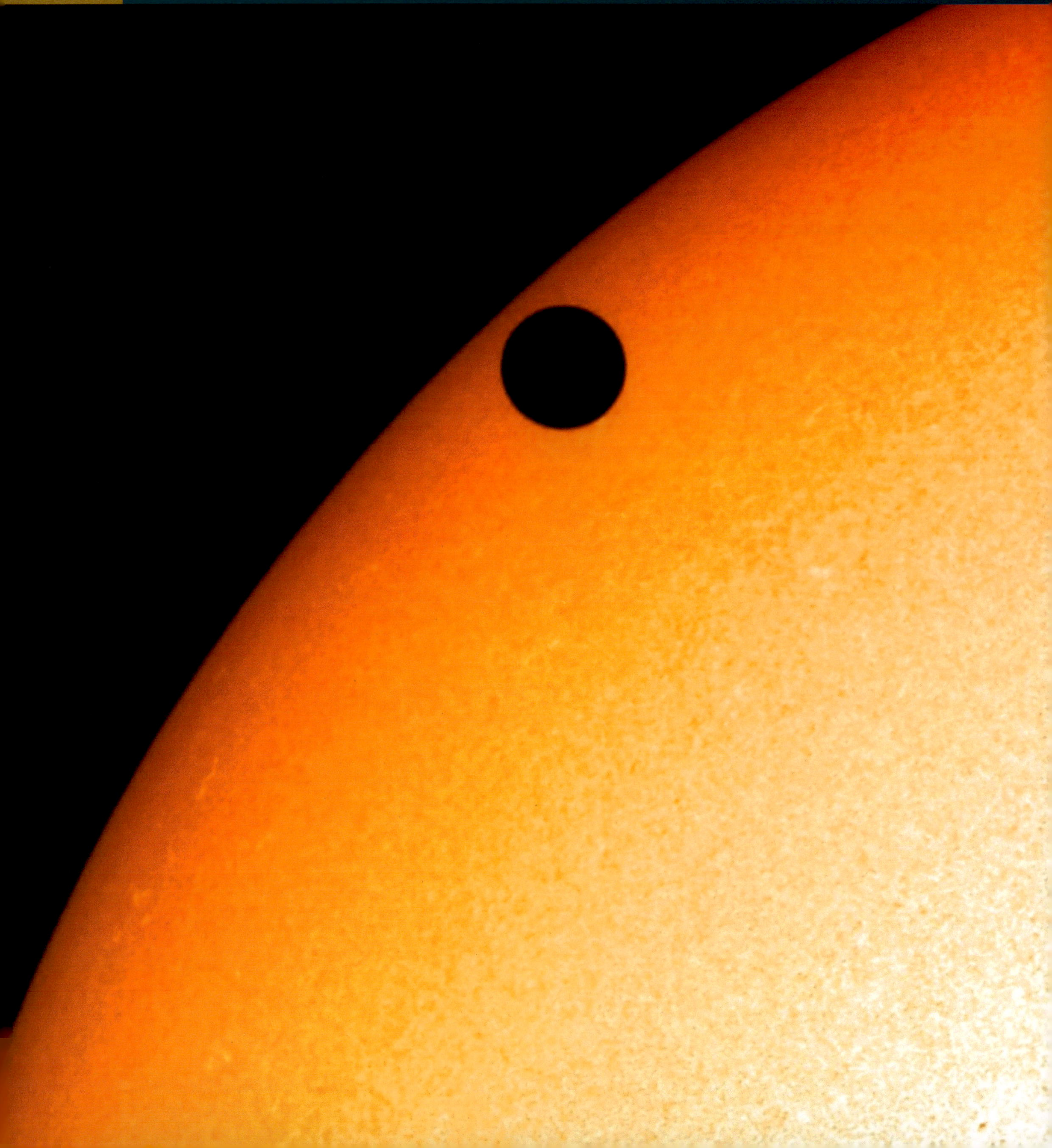

In 1761, Venus passed between Earth and the Sun. This was a rare event. Astronomers all over the world watched it. From different places on Earth, they noted when Venus passed in front of the Sun.

The next time Venus will pass between Earth and the Sun will be in 2117.

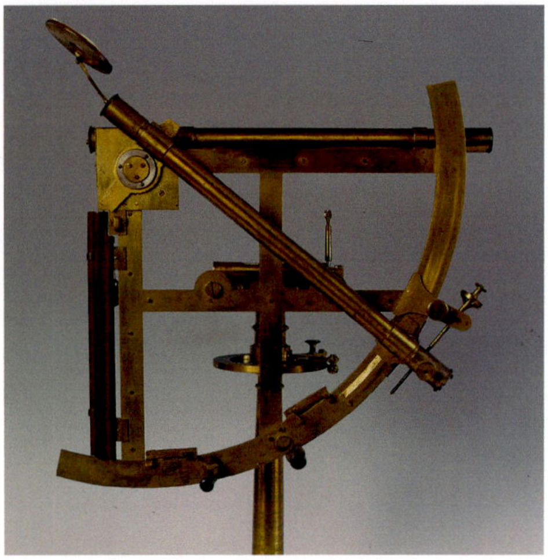

The astronomers measured the angles to Venus from different spots on Earth. They used the angle to figure out the distance to Venus. Once they knew the distance to Venus, they could figure out the distance to the Sun.

Light created inside the Sun takes thousands of years to reach its surface.

The northern lights are created by plasma from the Sun running into Earth's magnetic field.

Like Earth, the Sun has **magnetic poles**. It has a north and a south pole. Every 11 years, the poles switch places. Before they flip, the Sun becomes more active. It shoots plasma out into space.

magnetic field a protective layer around the Earth. It shields the planet against harmful particles in space

magnetic poles two small regions of a planet or star that are located near the north and south poles, toward which a compass needle points

All life on Earth depends on the Sun. Plants get energy from the Sun. Animals get energy from eating plants. The Sun keeps Earth warm. Without the Sun, everything on Earth would freeze and die.

A solar eclipse is when the Moon passes between Earth and the Sun.

Space Spotlight: Parker Solar Probe

NASA's *Parker Solar Probe* launched in 2018. Its job is to study the Sun. It is orbiting the Sun now. As it orbits, it gets very close to the Sun. It flies into the Sun's atmosphere. It flies through solar wind. The temperature is very hot there. The probe has a strong heat shield. The probe sends pictures back to Earth. After the probe runs out of fuel, it will burn up.

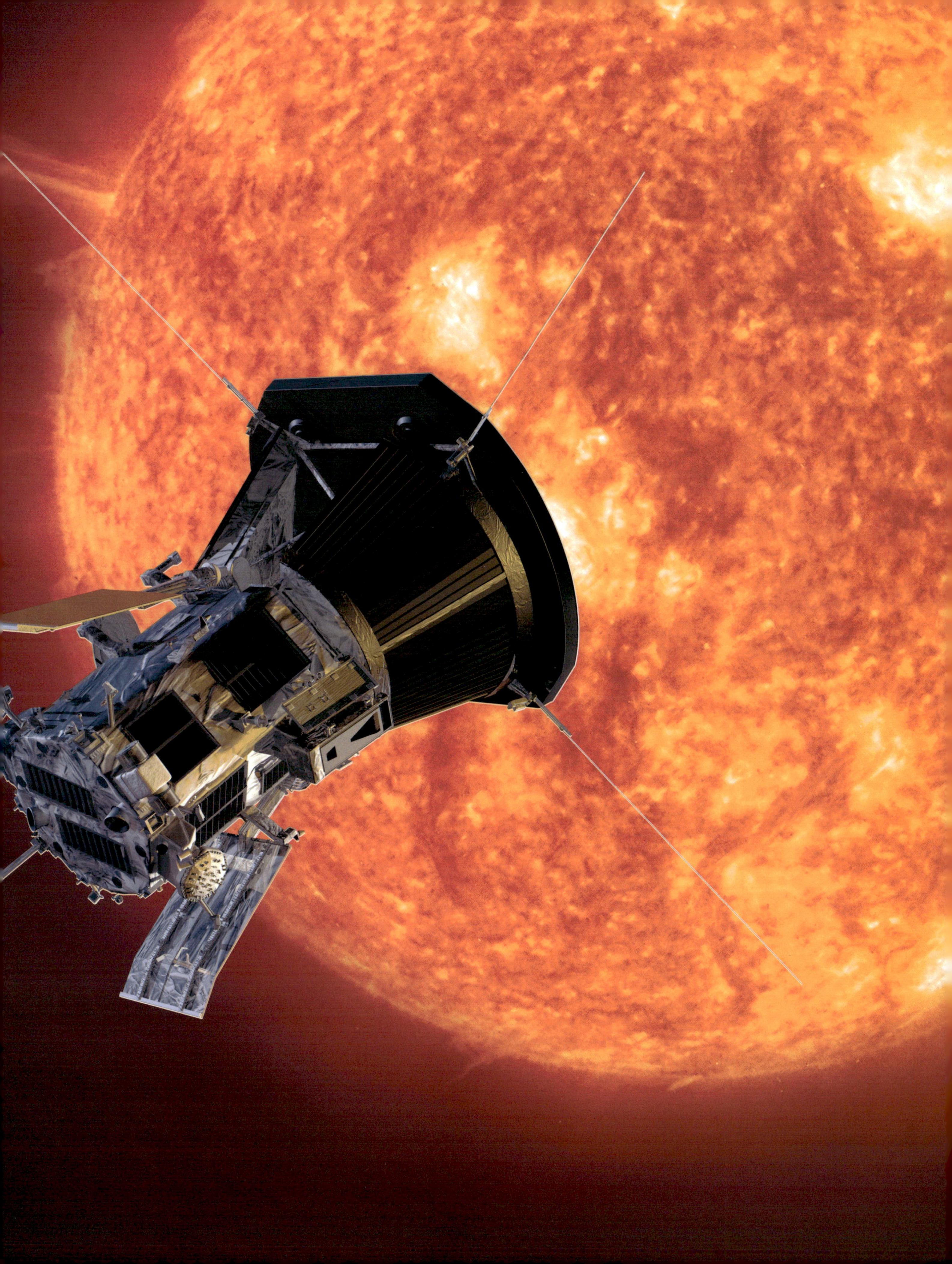

Read More

Bolte, Mari. *How the Sun will Save the World*. Minneapolis, Minn. Lerner Publications, 2022.

Huddleston, Emma. *Explore the Sun*. Minneapolis, Minn. ABDO Publishing, 2021.

Schuh, Mari. *The Sun: The Center of Our Solar System*. Minneapolis, Minn. Jump!, 2023.

Websites

NASA Science Space Place: Sun
https://spaceplace.nasa.gov/menu/sun/
Learn from NASA about the Sun.

Sun facts for kids
https://kids.kiddle.co/Sun
Learn more about the Sun.

Note: Every effort has been made to ensure that the websites listed above are suitable for children, that they have educational value, and that they contain no inappropriate material. However, because of the nature of the Internet, it is impossible to guarantee that these sites will remain active indefinitely or that their contents will not be altered.

Index